AF228153

PROTECTING OUR PLANET

POLLUTION

BY A. W. BUCKEY

Essential Library

An Imprint of Abdo Publishing
abdobooks.com

ABDOBOOKS.COM

Published by Abdo Publishing, a division of ABDO, PO Box 398166, Minneapolis, Minnesota 55439. Copyright © 2025 by Abdo Consulting Group, Inc. International copyrights reserved in all countries. No part of this book may be reproduced in any form without written permission from the publisher. Essential Library™ is a trademark and logo of Abdo Publishing.

Printed in the United States of America, North Mankato, Minnesota.

052024
092024

Cover Photo: Lane V. Erickson/Shutterstock Images
Interior Photos: Shutterstock Images, 4, 11, 15, 27, 43, 56, 68, 87, 100; Eric Paul Zamora/ The Fresno Bee/AP Images, 8; Gino Santa Maria/Shutterstock Images, 12; Sandusit Noom/ Shutterstock Images, 16; Natalia Kokhanova/Shutterstock Images, 18–19; Andreas Rentz/ Getty Images Entertainment/Getty Images, 24; Martyn Jandula/Shutterstock Images, 28; Ann Ronan Pictures/Print Collector/Hulton Archive/Getty Images, 31; RHS/AP Images, 35; Sean Gallup/Getty Images News/Getty Images, 36; John Gaps III/AP Images, 39; iStockphoto, 40, 92; Yasin Ozturk/Anadolu Agency/Getty Images, 46; Fokke Baarssen/Shutterstock Images, 48–49; Valentin Valkov/Shutterstock Images, 52; Mamunur Rashid/NurPhoto/Getty Images, 55; Antonio Cossio/picture-alliance/dpa/AP Images, 58–59; Silas Stein/dpa/picture alliance/Getty Images, 62; Donwilson Odhiambo/Getty Images News/Getty Images, 65; Mediacolors/Construction Photography/Avalon/Hulton Archive/Getty Images, 66; Vladimir Mulder/Shutterstock Images, 71; Gunter Marx/Alamy, 75; Vitaly Fedotov/Shutterstock Images, 76; Jeremy Chan/Getty Images Entertainment/Getty Images, 79; Roman Mikhailiuk/Shutterstock Images, 80; Thomas A. Jefferson/Blue Planet Archive, 84; The Ocean Cleanup, 90–91; Odua Images/Shutterstock Images, 97; Alex Wong/Getty Images News/Getty Images, 98; Petro Perutskyi/Shutterstock Images, 101

Editor: Haley Williams
Series Designer: Cynthia Della-Rovere

Library of Congress Control Number: 2023949548

PUBLISHER'S CATALOGING-IN-PUBLICATION DATA
Names: Buckey, A. W., author.
Title: Pollution / by A. W. Buckey
Description: Minneapolis, Minnesota: Abdo Publishing, 2025 | Series: Protecting our planet | Includes online resources and index.
Identifiers: ISBN 9781098293468 (lib. bdg.) | ISBN 9798384912736 (ebook)
Subjects: LCSH: Pollution--Juvenile literature. | Toxicology--Juvenile literature. | Air quality-- Management--Juvenile literature. | Radioactive pollution--Juvenile literature. | Water pollution control industry--Juvenile literature. | Air quality--Management--Juvenile literature. | Environmental sciences--Juvenile literature.
Classification: DDC 333.951--dc23

CONTENTS

CHAPTER 1
SAVING YOSEMITE 4

CHAPTER 2
THE SCIENCE OF POLLUTION 12

CHAPTER 3
THE HISTORY OF POLLUTION 28

CHAPTER 4
FOSSIL FUELS AND POLLUTION 40

CHAPTER 5
INDUSTRIAL AND AGRICULTURAL POLLUTION 52

CHAPTER 6
WASTE, SANITATION, AND POLLUTION 66

CHAPTER 7
WATER POLLUTION 80

CHAPTER 8
TAKING ACTION AGAINST POLLUTION 92

ESSENTIAL FACTS 100
GLOSSARY 102
ADDITIONAL RESOURCES....................... 104
SOURCE NOTES............................... 106
INDEX 110
ABOUT THE AUTHOR 112

Yosemite became the third US national park in 1890.

SAVING YOSEMITE

The previous night was a blur. Casey arrived at the airport in Fresno, California, around midnight, and Aunt Lainey greeted him with a big hug and a slice of cake. They drove in the dark to her cabin, and Casey could barely see anything of Yosemite National Park out the car window. Now it was morning, and Casey was ready to see the sequoia trees and the tall, squared-off rocks that he used to imagine were statues of giants.

Although he'd been to Yosemite many times, Casey still thought it was the most beautiful place in the world. When he was a kid, he visited the park for his birthday every July. But it had been years since he had been back. Aunt Lainey still lived in the same cabin and worked at the park as a geologist, studying the park's rocks and glaciers. For months, Casey had been dreaming about the hikes he would take with her through the park.

DANGER IN THE AIR

As Casey stepped outside the cabin, he saw Aunt Lainey on the porch with a cup of coffee. He walked over to join her and took a deep breath. The air outside still smelled of pine and fresh earth, like he remembered. But there was something else there too. It kind of smelled smoky. Casey turned to Lainey and asked if she was ready to get on the trail.

"I'm sorry, hon, but I don't think we should hike today," she said. "The air quality isn't looking too good." Lainey showed her phone to Casey. "We're at red today. That means the air is unhealthy. We'll be fine, but we probably shouldn't do any outdoor activities."

"What's wrong with the air? Was there an accident?" Casey asked. He was used to smoggy days in Houston, Texas, where he was from. The oil refineries there sometimes made the air smell like old eggs. But Casey knew Yosemite covered more than 1,100 square miles (2,849 sq km), with most of the land being wilderness.[1] He never thought of it as a place where the air quality could be bad.

"It's no accident," Lainey said. "We're just close to the Central Valley, one of the biggest farming centers in the country. When farmers burn their land, the fires release particles into the air. They're small enough to get inside your blood and can cause a lot of damage."

Casey was confused. "Why would farmers burn their own land? How does that help farming?"

"Farmers will sometimes burn their land to clear it for new crops," Lainey explained. "It's called agricultural burning. But that's not the only source of the fumes we experience. The Central Valley is home to a lot of cars and drivers. The exhaust from car engines ends up in the air and contributes to the smoke we see now. It also doesn't help that it's fire season."

"Fire season?" Casey asked. He didn't even know that could be a season.

"Here in California," Lainey said, "we have fire season. Summer to fall, the forests are hot and dry. If someone starts a fire by accident, there's plenty of tinder to help it get going, and it can spread faster than it can be contained. Fires may burn thousands of acres, and the smoke makes the air difficult to breathe. Just one cigarette left on the ground can cause a huge blaze.

One way wildfires can start is from people leaving their campfires unattended.

"All forests burn from time to time," Lainey continued. "Some fires are natural. But the climate in California keeps getting hotter and drier. It makes big fires more likely. In the last six years, we've seen more huge forest fires than ever before. And that means more smoky days like this one."

Lainey shook her head and smiled sadly. "Let me make you a good birthday breakfast," she said. "And if the air quality is better tomorrow, we'll take a hike out to the glaciers, OK?" Lainey stood

8

and headed back into the cabin. Casey took a long look at the wide, hazy sky and then followed her inside.

WAYS TO HELP

While Aunt Lainey was cooking, Casey opened his laptop and looked up *fire season*. He quickly discovered that what his aunt had told him was only scratching the surface of the problem. California's fire season had become longer, and air pollution had increased. As he read, Casey realized that fires were not just a California problem, but a global one. And car exhaust and other air pollutants Aunt Lainey told him about weren't just causing hazy days. They were also releasing gases that heat up the atmosphere.

As a result of human activity, Earth's climate was changing. Rising global temperatures were causing many forested places to become more dry. And wildfires are more likely to start in hotter, drier forests.

The more Casey researched, the more panicked he felt. Yosemite's land meant so much to him. The thought that a blaze could destroy the natural landscapes he loved was terrifying. Even more terrifying was the thought that the fires could harm people and places all around California and in the world. Casey wondered if there was something he could do to help.

At the end of one article about California's worsening fire crisis, Casey noticed a link to a website listing ways to help. He clicked the link and read about simple ways to help with

California forest fires. People could volunteer to help clear fallen leaves and other brush from forests. This is called fuel reduction, and it helps stop fires from getting bigger.

Casey liked that there were ways for everyday people to fight fires. But the problem seemed to go so much deeper than that. How could he help fight air pollution at the source?

After breakfast, Aunt Lainey offered to put on a movie, but Casey wanted to keep studying about ways to reduce air pollution. Lainey shrugged, and they sat in the den with their laptops. After some more research, Casey finally found what he was looking for. There was an organization back home called Air Alliance Houston that advocated for cleaner air. It fought for laws limiting harmful chemicals and was finding ways to lower the number of cars on the road. It also asked community members to volunteer and help measure air quality in different parts of the city.

Casey smiled. He'd never been that into science, but Aunt Lainey always made it seem kind of cool. She'd probably be interested in

Crops in the Central Valley

Changes in the climate of California's Central Valley can affect the world's food supply. The 18,000 square miles (46,600 sq km) of land produce about 25 percent of all food in the United States.[3] More than 300 different crops are grown there.[4] One famous California product is the almond. About 80 percent of the world's almonds come from the Central Valley.[5] However, a drier climate threatens almond crops because almonds require large amounts of water. One of the largest almond farms uses more water than every home in the city of Los Angeles, California.

a project measuring city air quality, and it made Casey want to take part. He wanted to learn more about what was in the air he breathed and why it made the planet warmer. Maybe he could even be a climate scientist one day.

Casey had a week left in Yosemite. If the air quality improved while he was there, he'd go hiking with Aunt Lainey. He would try to enjoy looking at the ancient sequoia trees and the waterfalls in the park. And then when he got back to Houston, the real work would begin.

The Environmental Protection Agency estimated around 66 million short tons (60 million metric tons) of air pollution was emitted in the United States in 2022.

THE SCIENCE OF POLLUTION

In the natural world, matter is always changing. Plants and animals die, and their remains become food for other life. Water flows from freshwater sources into the ocean and evaporates into the atmosphere. Environments adjust and change as living and nonliving things come and go. But sometimes a new substance, such as a pollutant, enters the environment and starts to affect organisms and the habitats they live in.

Pollution occurs when a substance or form of energy comes into an environment faster than it can be eliminated. For example, when a dead leaf falls on the forest floor, worms, fungi, and bacteria naturally consume it. The leaf is broken down, and any leftover material turns into gases such as carbon dioxide and nitrogen. This can be a quick or slow process, but the forest biome takes care of the change.

However, if a plastic bottle is left in the forest, it does not break down. Plastic cannot be eaten by worms or fungi because it does not decompose. Since plastic waste cannot be processed by the environment, it causes pollution.

Pollution can affect the world's air, water, and land. Matter that pollutes the environment, or pollutants, have a wide variety of impacts on human health and the natural world. Pollution has been a problem since the earliest human civilizations. Waste from ancient cities polluted nearby landscapes. Today, different types of pollution exist. Disposing of materials such as plastics and electronics presents new challenges.

Pollution introduces toxins into the environment. Toxins are substances that are harmful to living things. Different substances can affect living things in various ways. For example, caffeine, the stimulant in coffee, is safe for humans to consume in moderate amounts. But it is toxic to animals, such as dogs.

Toxins in the environment affect plants, animals, humans, and fungi differently. For this reason, the effects of pollution are very complex and difficult to fully understand. Predicting the long-term effects of pollution is also challenging. People may not notice or understand some consequences of pollution until years later. However, it is clear that pollution caused by human activities has increased over time. And it has already had a devastating effect on people and the planet.

Many communities recognize the threat of pollution. Across the world, scientists, lawmakers, activists, and citizens are coming together to fight pollution for the sake of cleaner air, water, and land. Fighting pollution may require people to make major changes, but everyone can take action to reduce the effects of pollution and create a cleaner future.

TYPES OF POLLUTION

Pollution can be classified in several ways. One way is to consider the source of the pollution. Pollution can come from cars, household items, power plants, and factories. Another way to classify pollution is by the pollutant itself. Lead pollution, for example, comes from many sources, including paint, aircraft exhaust, and factories. It can be found inside buildings and outside in water and soil.

Volcanic ash is considered a natural form of pollution.

Pollution can also be classified by considering the parts of the environment being polluted. This includes the land, air, and water. These areas of the environment can all affect one another. Pollution in the air can fall to the ground, and pollution on land can enter waterways.

However, pollution is not limited to substances that end up on land, in the air, and in the water. Sound and light can cause pollution as well. Noise pollution happens when sounds are so loud, persistent, or disruptive that they cause harm. Noise is measured in decibels, and a sound above 85 decibels can damage human hearing.[1]

Some loud noises occur naturally. For example, thunder is about 120 decibels.[2] But many human activities also create loud, ongoing noises. People who live near airports must deal with aircraft noise year-round, including at night. Studies from the

Civil Aviation Authority (CAA) in the United Kingdom found that people who live near airports are more likely to suffer from poor sleep and heart disease. The CAA also found that children who live near airports are more likely to have memory problems.

Like sound pollution, light pollution can disturb both humans and the natural world. One of the most widespread effects of light pollution is the glow from artificial lights kept on at night. Because of this glow, stars and other features of the night sky are less visible to humans than they once were. This is especially true in densely populated cities. A 2018 study found that elderly people living in places with a lot of artificial light were more likely to have trouble sleeping.

Light pollution also has harmful effects on ecosystems. Many bird species use the light of the moon as a guide at night, and artificial lights can distract or confuse them. According to the American Bird Conservatory, about four million birds die each year by running into artificially lighted structures.[3]

NATURAL VS. HUMAN-MADE POLLUTION

Pollution can happen naturally. Sometimes a natural event disrupts an environment and causes harm. A hurricane can cause water pollution by dumping seawater into brackish habitats, such as bayous. Sometimes natural pollution causes radical world changes. For example, the eruption of a volcano can pollute the air. In 1815, a volcano in Indonesia called Mount

Tambora erupted, sending ash and rocks into the air. The ash darkened the atmosphere worldwide. It also made the world colder. People called the year after the eruption "the year without a summer."[4]

Natural pollution is part of Earth's history. It is impossible to prevent completely. However, the recent increase in pollution worldwide is mostly due to human activity.

Sometimes, human activities can make some natural processes stronger and more intense. For example, carbon dioxide is a natural part of Earth's atmosphere. Humans and other animals breathe in oxygen and breathe out carbon dioxide. The world's plants then convert this carbon dioxide into oxygen.

The deep ocean also stores carbon from the atmosphere. The ocean's circulation carries some carbon into the water's depths. Other carbon is taken in by sea life that later dies and sinks to the seafloor. The process of releasing and storing carbon is called the carbon cycle. However, human inventions, such as combustion engines, release more carbon than this cycle can process. When this happens, the amount of carbon in the atmosphere will then increase, causing pollution.

In other cases, human activities create new pollutants not found in nature. These are synthetic chemicals or materials. One example is plastic. *Plastic* is a word to describe something

Fertilizers and Pollution

Plants need nutrients to grow. Farmers and gardeners want to raise the tallest, healthiest crops possible. Fertilizers help by supplying more nutrients than are naturally found in soil. The three most common fertilizer ingredients are nitrogen, phosphorus, and potassium. Although these fertilizers can help plants grow, they can also pollute waterways. Nitrogen and phosphorus can cause algae to overgrow in bodies of water, decreasing water oxygen levels. Nitrogen compounds called nitrates can pollute drinking water and cause harm to those who drink it. A common potassium fertilizer can make water saltier, which can harm the ecosystem.

that bends or stretches easily. Today, people use the term to mean mostly synthetic polymers, which are materials with a bendable quality.

The first synthetic plastics were invented in the early 1900s. Plastic quickly became popular for its flexibility. It could be processed into many shapes, giving it a wide range of uses. And unlike some natural materials such as wood, plastic took little time to create and process. Vinyl, Styrofoam, and Plexiglas are all types of synthetic plastic. Some plastics are made of natural materials such as cellulose, or plant fibers. But many plastics come from oil, natural gas, and coal. Synthetic polymers are light and easy to produce. Plastic is especially popular as a packaging material. It is effective at sealing goods, and its light weight helps companies save money on transport costs.

Once created, plastic does not break down easily. Materials that are biodegradable can break down with the help of organisms and sometimes oxygen. Paper, for example, is made from plants and is biodegradable. Plastic, on the other hand, is not biodegradable. Once plastic is used, it stays in its original form for many years. A thin plastic bag can break down in about 20 years. A plastic bottle takes more than 400 years.[5]

Even when plastic does break down, it does not transform into something else. When composted, food waste such as banana peels decompose into humus, a rich material found in soil. But plastic decomposes into smaller and smaller pieces of plastic. The natural world has no way to process those tiny pieces, meaning they stay there and pollute the environment.

Synthetic chemicals are also useful for human activities, but they can easily become pollutants. Pesticides are substances that kill animals that are considered pests. In buildings, people use pesticides to kill animals such as cockroaches and mice. Farmers put pesticides on crops to protect them from insects that might eat them. An effective pesticide can save important crops from destruction. But like synthetic plastics, synthetic chemicals do not break down quickly. These chemicals can stay in soil for many years, continuing to kill wildlife long after serving their intended purpose. They can also make their way into bodies of water such as rivers and lakes.

COMMON POLLUTANTS

Any substance can be a pollutant if it disrupts the environment and causes harm. But some pollutants are more widespread than others. Air pollutants tend to be very tiny solids or liquids capable of staying suspended in air. These pollutants are called particulate matter. Other air pollutants are gas molecules that become part of the atmosphere. Ozone, carbon monoxide, and nitrogen oxide are some of the most common. These air pollutants have

many effects, but all of them can cause breathing problems
when inhaled.

Many kinds of substances can dissolve in water. As a result,
water is easily polluted. Some of the most common water
pollutants include pesticides, fertilizers, wastewater, oil, bacteria,
and heavy metals such as lead and mercury. Water can also be
contaminated by radioactive elements. These elements give
off a harmful form of energy. Some radioactivity is natural in
low doses. Metals such as
uranium are radioactive and
found in water. At higher
doses, though, radiation can
damage the cells of living
things. Nuclear power plants
can be sources of radioactive
water pollution.

Land and water are
closely linked. Waterways
and the surrounding
landscape can exchange
pollutants, meaning
many pollutants are found
in both places. Common

Superfund Sites

Superfund sites are places in the United
States where hazardous waste was left
behind or handled incorrectly. As a result,
these places are dangerous for people to
live near. One Superfund site is a former
paint store in Milford, New Hampshire,
called Fletcher's Paint Works. In 1982, state
officials found leaking containers of paint
chemicals stored there. These chemicals
had entered the water supply and soil in
the area. The EPA works to clean up the
aftereffects of Superfund sites and to hold
the responsible companies accountable for
their actions.

land pollutants include heavy metals, pharmaceuticals, and
plastics. Garbage from human activity, both inside and outside
landfills, also pollutes Earth. Some types of pollution can make

MELATI AND ISABEL WIJSEN

Melati and Isabel Wijsen grew up on the island of Bali in Indonesia. Bali is a famous tourist destination known for its beautiful beaches. When the sisters were younger, they became sad and angry over the amount of used plastic they saw on the island.

Melati was 12 and Isabel was 10 when they decided to take action. Together, the sisters started a campaign to ban single-use plastic on the island. Their work began in 2013. By 2019, they had achieved their goal. The governor of Bali banned plastic bags, plastic straws, and Styrofoam.

Today, the Wijsen sisters run a nonprofit called Bye Bye Plastic Bags (BBPB). They create educational materials to teach children about the dangers of plastic pollution. BBPB also participates in river cleanups. It continues to influence governments to enact plastic bag bans. The organization has more than 50 branches worldwide. The sisters hope to encourage other teens to do similar work to fight pollution. Melati said, "We're a living example that kids can do things."[7]

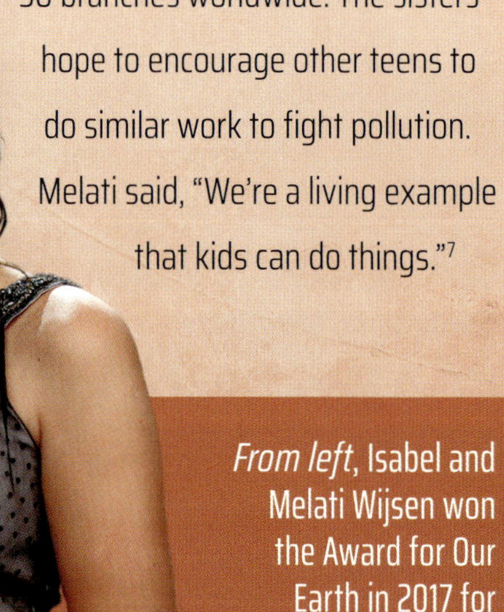

From left, Isabel and Melati Wijsen won the Award for Our Earth in 2017 for their contributions to nature conservation.

land unsafe for long periods of time. Brownfields, for example, are areas of land that cannot be fully used because they are too polluted. The Environmental Protection Agency (EPA) estimates there are about 450,000 brownfields across the United States.[8]

SHORT- AND LONG-TERM EFFECTS OF POLLUTION

The effects of pollution can show themselves very quickly. In humans, air and water pollution can lead to serious illnesses and sometimes death. Even short-term, limited exposure to polluted air or water can cause harm.

In June 2023, a series of wildfires in Canada released smoke and fumes that polluted several US states. On June 7, 2023, the Air Quality Index (AQI) in New York City was higher than 400.[9] An AQI level below 100 is considered good air quality, while anything above that is unhealthy. An AQI between 301 and 500, which is the highest end of the index, is considered hazardous for people's health. When air pollution levels are that high, people with health conditions such as heart disease can risk illness simply by going outside.

The effects of water pollution can also be dangerous in the short term. Globally, more than one million people each year die as a result of unsafe drinking water.[10] Many of the harmful contaminants in drinking water come from human activities.

Chemicals from pesticides and fertilizers cause pollution when rain washes them into nearby bodies of water and is then consumed. Sewage, which is waste transported through sewers, is a significant source of disease-carrying bacteria and parasites. It can cause many illnesses if it gets into drinking water. Land pollution can also impact human health when pollutants such as pesticides enter the food supply through the soil.

Scientists still don't completely understand the effects of pollution over a longer period. For example, as plastic breaks down into smaller pieces, it creates tiny particles known as microplastics. Microplastics are now widespread in soil, water, and air. They are even inside living things.

In 2022, scientists tested blood donated by 22 adults and found microplastics in the bloodstream of 17 of them.[12] But scientists are uncertain about the effects of plastics in the bloodstream. When it comes to pollution, there are many unknowns. While it has always been a part of human history, the types and severity of pollution have changed significantly in the last few centuries.

Some effects of pollution can be addressed with short-term solutions. For example, metals such as iron and manganese can pollute water. However, removing these metals can be done somewhat easily. Filters and water softeners, which replace metals with sodium ions, both effectively treat metal pollution.

Other types of pollution are more difficult to manage. Air cannot be contained in the same way that water can, making

air pollution impossible to truly contain. Respiratory masks help filter out harmful air particles and prevent people from inhaling them. Indoor air filters and purifiers serve a similar purpose. But once pollutants are released into the atmosphere, there is no way to filter and remove them on a larger scale. Because the effects of pollution can be so difficult to reverse, many approaches to air pollution focus on preventing or minimizing it.

Smog is a type of air pollution
that is common in big cities.

THE HISTORY OF POLLUTION

Pollution is not a new problem. Fires lit by early humans were some of the first human-related sources of air pollution. Along with that, people tend to live in groups, and groups of people create waste. Early humans often piled their waste in trash heaps. These heaps of waste would decompose more slowly than other natural remains, releasing heat and gas. Archaeologists have found these trash piles, called middens, all over the world. Some are many thousands of years old.

EARLY POLLUTION SIGNS

As the human population grew, so did its impact on the surrounding environment. The ancient Romans were a culture based in what is now Italy. By the first century BCE, the Romans expanded their society into an empire that covered much of Europe, North Africa, and some of the Middle East. The Romans mined metals for things such as water pipes and tools.

Studying the air pollution that existed in ancient times may seem impossible. But scientists have found a way by analyzing ice in glaciers. Air from ancient times is trapped inside the layers of glacier ice. Scientists can study this air to determine what pollutants were in the air during ancient Roman times. They have found that the Romans' mining and metal production over a 500-year period left toxic waste in the air, especially lead fumes.

In the past, some people believed air pollution was bad for health. However, these people didn't always understand the connection between garbage, water and soil pollution, and illnesses.

For example, cholera is a disease that can cause severe diarrhea and dehydration. It is caused by bacteria that survive inside an infected person's feces. If people drink water that has been exposed to human waste, cholera can spread very quickly. And those left untreated could die from it.

By the 1800s, people were beginning to understand that very small living things could cause disease. In 1854, English doctor John Snow helped prove that contaminated drinking water and cholera were connected. Over time, people realized that keeping water clean was key for human health.

THE INDUSTRIAL REVOLUTION

Beginning in the 1700s, machinery made the production of goods faster and cheaper than before. New materials and energy sources were used in factories, and cities grew as industries expanded. This era, which lasted into the early 1900s, became known as the Industrial Revolution.

The Industrial Revolution led to a huge increase in global populations, leading to higher amounts of pollution being produced. It also caused a major change in the organization of

Experts believe that crowded living conditions and poor sanitation practices were a big factor in the spread of cholera and other diseases during the 1800s.

global societies. Before the 1700s, most societies made goods by hand. Farms were at the heart of food and goods production. Farmers grew food and raised animals for products including cloth and leather. Other manufacturing such as metallurgy happened on a small scale.

But during the Industrial Revolution, machines were invented to make production easier. For example, an invention called the power loom allowed people to weave cloth automatically instead of by hand. Cloth could be made more quickly than ever before. This invention and others like it changed the way people worked. Instead of making things at home, many people moved to cities to work in factories that produced goods at a faster rate.

The Industrial Revolution began in Europe and quickly spread to other parts of the world. It also helped produce a population boom. The world's population began to increase at greater speed. Between 1625 and 1825, the global population doubled from 500 million to one billion. It took only about a century to double again, reaching two billion by 1930.[2] More people meant more goods being produced, more travel, and more crowded cities.

The Industrial Revolution was also a revolution in industrial pollution. During this time, people learned to harness electricity, and they wanted this new form of energy to light their homes. Electricity was generated by power plants fueled by coal-burning engines. These engines also helped power the factories that drove the Industrial Revolution. As a result, the release of smoke and other air pollutants skyrocketed.

In many places, new fuels and industries brought a decline in air quality that was impossible to ignore. Manchester, England, was one of the first modern industrial cities. In the 1800s, it became famous for its dirty, dark skies and the breathing problems its people suffered.

In the 1850s, English scientist Robert Angus Smith found that due to the city's air pollution, rain in Manchester contained acidic compounds that harmed people and the environment. This phenomenon, known as acid rain, would not be widely studied until the 1900s. However, it was already present in heavily polluted places.

The first automobiles were invented in the late 1800s. In the early 1900s, US businessman Henry Ford realized he could use factory workers to make cars quickly. Rather than assembling one car at a time, Ford's company divided the work of building a car into many small tasks done by many workers on an assembly line.

Ford's innovation made the automobile cheaper and more accessible. By the 1920s, cars were common in the United States. An increase in car ownership meant an increase in air pollution from burning gasoline, the fuel used to run cars. Between 1900 and 1950, the carbon emissions per person in the United States had almost doubled.[4]

THE BEGINNING OF THE ENVIRONMENTAL MOVEMENT

People have been concerned about pollution for many centuries. Ancient civilizations looked for ways to keep soil healthy. They worried about the effects of dirty air. But some of the more modern concerns about industrial pollution date to the Industrial Revolution. For example, US founding father Benjamin Franklin worked to remove tanneries, or leathermaking businesses, from his home city of Philadelphia, Pennsylvania. He argued that citizens had a right to clean air and the tanneries dirtied it.

The 1900s brought many new technological innovations. Concerns about the effects of new inventions helped create the modern environmental movement. In the 1940s, scientists developed an insecticide called DDT. It was very effective at killing mosquitos and other insects that spread disease and killed crops. However, DDT had harmful effects on people and wildlife.

Rachel Carson wrote several books during her lifetime, including *The Sea Around Us*, which discusses the importance of Earth's oceans.

In 1962, scientist and writer Rachel Carson released her book *Silent Spring*. The book explained that DDT did not disappear after being sprayed on crops or in the air. Instead it stayed in human and animal tissue. In addition, DDT entered the water supply. The chemical caused illness in humans and seemed to increase risk of cancer. It also interfered with wildlife reproduction.

The popularity of *Silent Spring* eventually led to the ban of DDT in the United States. It also raised awareness of the need to regulate pesticides and other chemicals released into the environment. Around this time, environmentalism became more popular. People came together to protect forests, fight air pollution, and combat harmful chemicals in soil and water.

In response to the rise of the environmental movement, the US government created agencies to address environmental problems. In 1970, the government created the EPA. The agency's job is to promote clean air, water, and land in the United States. The EPA studies environmental issues and helps write and enforce environmental laws. Some of the most important antipollution

Many towns near Chernobyl, Ukraine, had to be evacuated after the explosion. Today, the city of Pripyat remains abandoned.

laws are the Clean Air Act of 1970 and the Clean Water Act of 1972. These laws regulate pollutants entering the air and water.

GLOBAL CRISES AND EFFORTS TO ADDRESS POLLUTION

In the late 1900s, several famous disasters increased people's awareness of pollution's risks. The Chernobyl nuclear power plant operated in Chernobyl, Ukraine. Nuclear power plants work by splitting atoms and harnessing the energy that is released. They do not release air pollution in the way that coal power plants do, but they do create radioactive waste. And if something goes wrong, radioactive material can be released into the surrounding environment.

In 1986, this happened at Chernobyl. An explosion destroyed part of the power plant, and radioactive material was released. The town near the power plant had to be evacuated.

Some plant workers and rescue personnel died from exposure to intense radioactivity near the plant. Lower levels of radioactive energy spread to countries as far away as France. The radiation from the power plant caused long-term damage to the forest ecosystems in the area. Today, the area is still not inhabited by humans.

> If we are going to live so intimately with chemicals—eating and drinking them, taking them into the very marrow of our bones—we had better know something about their power.[5]
>
> —Rachel Carson, *Silent Spring, 1962*

By the late 1980s, scientists were aware of another effect of air pollution. They knew that burning fossil fuels such as coal and oil released carbon dioxide and other gases into the atmosphere. They also knew that emissions of these gases had increased significantly since the Industrial Revolution. These emissions are known as greenhouse gases because they trap heat in the atmosphere.

It became clear during this time that greenhouse gases were building up in Earth's atmosphere and the planet's average temperature was rising. People became aware of the link between pollution and hotter temperatures around the world. This connection is called global warming. It is one part of a broader shift in Earth's climate, known as climate change. In 1988, the United Nations (UN) established the International Panel on Climate Change (IPCC). The IPCC was tasked with collecting information on climate change and figuring out the effect it had on the world.

Other dangers of the oil industry also became

apparent in the late 1980s. In 1989, the oil tanker *Exxon Valdez* got stuck in shallow water off the coast of Alaska. It spilled 11 million gallons (42 million L) of oil into the ocean. The oil killed billions of fish eggs and about 250,000 birds, as well as other marine species. The 1,300 miles (2,100 km) of affected shoreline never completely recovered.[6] The *Exxon Valdez* spill serves as a reminder of the risks of oil production and transportation.

A pump jack is a device used to extract oil from underground wells.

FOSSIL FUELS AND POLLUTION

Fossil fuels get their name from their source. A fossil is a trace of a long-dead life-form. Dinosaur bones that have turned into rock are probably the most well-known examples of fossils. Fossil fuels are the remnants of dead, ancient organisms. They form from the decomposition of plants and animals.

Fossil fuels are found inside Earth's crust. The major fossil fuels are oil, natural gas, and coal. These fuels are hydrocarbons, which means they are made mostly of the elements hydrogen and carbon. All three are used in important energy processes, such as heating homes and making engines work. But the use of these fuels releases much of the world's most harmful pollution.

TYPES OF FOSSIL FUELS

Oil is a fossil fuel found in cracks between rocks in Earth's crust. It formed when tiny sea organisms called diatoms died and their remains were trapped underground. Like all life-forms, diatoms were organic, meaning their bodies contained carbon as an essential building block. Over millions of years, heat and pressure transformed these organisms into a carbon-rich liquid called crude oil, or petroleum.

Crude oil is extracted from underground reserves. These reserves are found in places that once formed ancient ocean beds hundreds of millions of years ago. Crude oil deposits are found all over the world, but there are especially large reserves in certain areas.

A little more than two-thirds of the world's crude oil is in the Middle East.[1] A large number of oil reserves can be found in Saudi Arabia, Iran, and Iraq. But the country with the most reserves is Venezuela. People must find the oil reserves before they can be used. Once the reserves are identified, workers drill down into the land or seabed. They extract the crude oil with pumps.

Crude oil has many uses, including as a raw material to make plastic. However, it's typically used as the main ingredient in the gasoline that fuels cars and other machines. Inside a car's internal combustion engine, gasoline is mixed with air and ignited to create a tiny explosion, or combustion. The engine turns this

In 2022, cars and vans emitted about 3.9 billion short tons (3.5 billion metric tons) of carbon dioxide.

energy into a rotating motion that spins the car's wheels. This process also releases gases such as carbon monoxide, nitrogen oxides, and hydrocarbons. These gases are air pollutants.

Similar to crude oil, natural gas forms from decomposed matter underground. While pressure turns some of this matter into the liquid that makes oil, sometimes pressure and heat transform it into gas. Natural gas is found in underground reserves and is then extracted. It is used mostly to make power for electricity and to heat people's homes. Russia has the largest natural gas reserves in the world, followed by Iran. The United States is one of the top five countries with the most natural gas reserves, holding 6.5 percent of global reserves in 2021.[2] In 2022, approximately 40 percent of US energy came from natural gas.[3]

Coal is a fossil fuel found in deposits in the ground. It burns easily and is made mostly of carbon. People have used coal for centuries, but its use became more widespread during the Industrial Revolution. Burning coal releases heat, which can be turned into another form of energy. For example, a coal power plant burns coal to heat water into steam, and the steam spins a turbine. A device called a generator turns the spinning motion of the turbine into electricity. One-third of the world's electricity is created this way.[4] Coal is also used to heat homes.

Coal contains more carbon than other fossil fuels, so it releases more carbon into the atmosphere when burned. Coal mining and processing also cause water pollution. Some coal must be mined from deep inside Earth. The mining process releases harmful substances into nearby water sources.

Coal mined above ground pollutes the land. In addition, coal is washed after it is mined. The washing process creates a by-product called coal slurry. This black sludge can pollute rivers, lakes, and streams.

Fracking

Because fossil fuels are often buried deep underground, they can be difficult to extract. One method of fossil fuel extraction is called hydraulic fracturing. It is commonly known as fracking. In this process, a machine injects pressurized liquid mixed with sand into underground spaces. The liquid breaks open rocks, helping release natural gas. This process creates a few types of pollution. The injected liquid has chemicals in it and becomes wastewater. The fracking process causes air pollution. It is also loud, raising noise pollution concerns.

Even though fossil fuels are made from natural processes, they are nonrenewable. This is because the process of converting a living thing into a fossil fuel takes a very long time. Humans are on track to use up Earth's fossil fuels long before new supplies of coal, oil, or natural gas will be naturally produced again. No one knows exactly when fossil fuels will be gone. But based on known reserves and current usage levels, experts estimated in 2020 that oil and natural gas may last about 50 more years. And there may be enough coal to last 140 years.[5]

EFFECTS OF BURNING AND PROCESSING FOSSIL FUELS

When fossil fuels are burned, they release airborne particles of carbon known as soot. They also release the greenhouse gases carbon dioxide and nitrous oxide. These are the gases that contribute greatly to global warming. In addition, nitrogen oxides released by burning fossil fuels cause smog and acid rain. All these pollutants from fossil fuels can have widespread effects.

However, some communities are more deeply affected by fossil fuels than others. The people who live on Navajo Nation land near Upper Fruitland, New Mexico, are one example. Their homes are near a coal mine and a coal-powered electric plant. This plant and a nearby one in Arizona emit large amounts of nitrous oxide.

The people who live near Upper Fruitland feel the effects of pollution daily. Those who live downwind of the plant report that their homes are regularly covered with black dust. The residents say that their community has especially high rates of asthma and bronchitis, diseases that affect the lungs. They believe the plant is to blame. Long-term residents told the *Los Angeles Times* that they remember a time when there were more plants and animals on the land. Today, they say that wildlife is gone.

Some people protest for governments to stop using oil or gas pipelines that run through important American Indian tribal lands.

THE SEARCH FOR FOSSIL FUEL ALTERNATIVES

In many ways, fossil fuels have helped power the world. But the environmental costs of using fossil fuels are high. Many people are looking for alternate ways to heat homes, power cars, and keep electricity flowing. They want to use energy sources that are renewable and produce no harmful emissions. A 2022 survey found that 69 percent of US residents believed that the country should make finding renewable energy sources a priority.[7]

Alternatives to fossil fuels are older than many people realize. Early electric cars first appeared in the late 1800s. But at the time, these vehicles were impractical. Gasoline-powered cars were easier and cheaper to make. Today, electric car technology has come a long way. By 2023, more than 40 types of fully electric cars were on the market.[8] These cars are charged by plugging them into an outlet just like a phone or a lamp.

> The air is unbreathable. The heat is unbearable. And the level of fossil fuel profits and climate inaction is unacceptable. . . . We have seen some progress. . . . But none of this is going far enough or fast enough.[9]
>
> —António Guterres, UN Secretary-General, 2023

They do not produce emissions when driving. However, if they are charged using electricity generated by fossil fuels, they still contribute to fossil fuel emissions.

Besides fossil fuels, there are other options to produce energy. For example, solar panels can turn the sun's light into electricity.

Wind farms are sometimes built in the ocean because wind speeds are usually faster over water than they are over land.

The panels do not emit air pollutants as they generate electricity. Solar power has become an important renewable energy source worldwide. In 2022, almost 5 percent of US electricity came from solar energy.[10]

Wind energy is another renewable energy source that does not emit greenhouse gases. To create wind energy, wind turns the blades of a wind turbine, and then a generator turns that motion into electricity. As with solar power, generating wind energy does not produce emissions, and because of that, it has also

become an important renewable energy source. Approximately 10 percent of the electricity in the United States came from wind energy in 2022.[11]

Nuclear energy is another renewable energy source. It works by splitting atoms in a controlled way. This process releases large amounts of energy in the form of heat. Heat is used to produce steam, which spins a turbine. A generator turns that spinning motion into electricity. In 2022, around 18 percent of US electric power was generated by nuclear energy.[12] Nuclear energy does not emit greenhouse gases, and it is a highly efficient source of power. However, nuclear plants must be carefully monitored to ensure that they do not leak radioactive material into the environment.

By switching to renewable and clean energy sources, humans could significantly reduce fossil fuel pollution. But making a switch from fossil fuels to other energy sources can be costly in the short term. Changing energy sources could mean reorganizing the way the economy works. For example, the residents of Upper Fruitland are harmed by their local coal power plant. But the plant also provides well-paying jobs to community members. For this reason, many are reluctant to ask the government to shut down the power plant.

Some governments around the world created programs to make it easier for people and businesses to make the switch to alternative energy. For example, the US government offers tax credits to people who choose to install solar panels in their homes. Businesses can get similar tax credits from the government. In addition, the international community has created goals for reducing greenhouse gas emissions.

One of the most recent international protocols for addressing the linked threats of climate change and pollution is the Paris Agreement. The Paris Agreement was created at the 2015 UN Climate Change Conference and was ratified in 2016. It was signed by 196 parties, the vast majority of the world's countries. The goal of the Paris Agreement is to commit to action that will reduce greenhouse gas emissions and limit the rise of global temperatures. Each country has different goals for reducing emissions, and every five years, countries increase their goals to help continue decreasing emissions.

In addition, the Paris Agreement asks for developed countries to help developing countries invest in climate change solutions. Each year, the UN hosts a Climate Change Conference, known as the Convention of the Parties (COP). This conference offers a chance to see where each country is at in its goal to reduce emissions.

The COP also invites activists and speakers, including young activists, to share their views on pollution and climate change. In 2023, youth climate activist Alab Mirasol from the Philippines and Pakistani activist Ayisha Siddiqa, who founded an environmental organization called Polluters Out, were both invited to speak at COP27 in Sharm el-Sheikh, Egypt. They along with many other activists around the world work to address the threat of pollution.

The Kyoto Protocol

The Kyoto Protocol was the first international treaty working toward reducing global greenhouse gas emissions. It was created in 1997 and ratified in 2005. Although the Kyoto Protocol is still in effect, it was superseded by the Paris Agreement. The purpose of the Kyoto Protocol was for the 192 countries involved to individually reduce their greenhouse gas emissions by a certain amount. Since many developed countries are some of the biggest contributors to greenhouse gas emissions, the Kyoto Protocol targeted those countries to reduce their emissions by 5 percent based on their 1990 levels.[13] Although the United States originally signed the Kyoto Protocol, it never actually ratified it.

Approximately 52 percent of land in the United States is used for agriculture.

INDUSTRIAL AND AGRICULTURAL POLLUTION

Individual people contribute to pollution over the course of daily life. Driving a car, littering, or even doing laundry can pollute the air, land, and water. But today, much of the world's pollution comes from businesses and industries.

Industrial manufacturing and transporting goods are major contributors to greenhouse gas emissions and water pollution. Agriculture provides people with food and essential goods, such as cloth. However, it is also one of the top contributors of global emissions. Industrial and agricultural pollution continue to threaten communities, wildlife, and ecosystems around the world.

INDUSTRIAL POLLUTION

Industrial pollution comes from facilities that produce materials and goods. The factories that make goods are often powered by fossil fuels. In fact, in the United States, 23 percent of greenhouse gas emissions come from industry.[1]

Manufacturing is a general term for the creation of new goods. Most manufacturing creates waste. This waste can enter the environment in the form of land, air, and water pollution. Some products and processes are more likely to pollute than others.

For example, tanning is the process of manufacturing leather from animal skins. It uses chemicals to soften and change the skins. The skins are then soaked, and the water used to soak them becomes wastewater, which may contain dangerous chemicals. The substances used to tan leather also pollute the air. The arsenic used in the tanning process can enter the air as particles, harming workers' lungs.

Water that is not absorbed by the land is called runoff. Industrial runoff from manufacturing sources often creates pollution. This runoff can come from point sources, meaning that wastewater is dumped directly into a water source at a specific place. It can also come from indirect sources. This happens when pollutants enter the soil and rainfall or irrigation eventually carries the pollutants into bodies of water.

Pollution and Global Inequality

Wealth and discrimination play a role in exposure to pollution. A 2019 study in the *International Journal of Environmental Research and Public Health* found that across countries in Europe, lower-income communities were more likely to be exposed to air pollution. A study in the same journal found a similar link between income and pollution exposure in China. And across industries, people who work in factories or on farms must often endure unhealthy working conditions in order to survive. In Indonesia, many factory workers have reported that long-term breathing issues and even death are common at their workplaces.

Toxic wastewater from textile factories in Bangladesh has severely polluted many of the country's rivers.

To avoid polluting the environment with industrial waste, manufacturers must treat, or clean, their waste before releasing it. Manufacturers can build their own water treatment plants to deal with waste before it becomes runoff. They can also send waste to local water treatment facilities, such as sewage plants. However, treating water is expensive. Irresponsible manufacturers may try to get rid of wastewater without treating it properly.

Wolverine Worldwide is a company that makes shoes in Rockford, Michigan. The company grew quickly after its soft leather shoes, Hush Puppies, became a hit with customers in the 1960s. Wolverine's factories brought wealth and growth to the town. But rapid growth led to an increase in pollutants.

The company tried to get rid of waste from its tannery in various ways. It dumped waste at local sites and tried to sell it to farmers. Eventually, a government report found that Wolverine may have disposed of waste in more than 100 places near Rockford. And without proper treatment, harmful chemicals found their way into water.

In the late 2010s, the water near Rockford tested positive for substances known as PFAS. These synthetic chemicals are often used as coatings or protectants, such as for cookware. They are considered hazardous. PFAS have been linked to increased risk of cancer and reproductive problems in humans. They also have similar effects on animals. Scientists have found links between PFAS exposure and health problems such as lower fertility and slower healing in some animals.

Pollution can cause lower oxygen levels in bodies of water. When oxygen levels are too low, animals and plants will die.

INDUSTRIAL POLLUTION IN THE FASHION INDUSTRY

The fashion industry is one of the world's top pollution-causing industries. Clothing manufacturing is a huge global business. About 100 billion items of clothing are produced each year.[2] Most of that clothing is made in China, a country that accounts for 28 percent of all manufacturing worldwide.[3]

Creating clothing is a complex, multistep process. Raw materials such as cotton for cloth or cowhides for leather must be grown, collected, harvested, and transported, resulting in carbon emissions. Clothing factories burn fossil fuels and produce industrial runoff. And discarded scraps and unsold clothing create waste that can pollute Earth. According to the UN, about 10 percent of global carbon emissions are produced by the fashion industry.[4]

For example, many companies dump unwanted clothing in the Atacama Desert in Chile. This desert is near an international port, where clothes are shipped to be resold. However, much

of the clothing sent to Chile is not sold or accepted by landfills. Instead, the clothing is left in the desert. The clothing then forms "dunes" in the desert, covering the soil and preventing plant growth.

The fashion industry is a large consumer of plastic as well. About 60 percent of materials used in the fashion industry are made from plastic.[5] The industry is also a major source of pollution in waterways. Producing clothes requires a huge amount of water. Cotton, the most common clothing fabric, is

a water-intensive crop. Turning cotton into clothing requires washing, dyeing, and sometimes bleaching it.

During the manufacturing process of turning raw materials into textiles, around 8,000 synthetic chemicals are used.[6] These chemicals then end up in the wastewater from the clothing production. If that wastewater is dumped into waterways, the chemicals are released into the water. Harmful chemicals such as lead, mercury, and arsenic can lower the quality of drinking water and can cause health issues in both humans and wildlife.

About one-fifth of the world's industrial wastewater pollution comes from the fashion industry.[7] This pollution is widespread, especially in areas where manufacturing is common. In China, about 70 percent of rivers are polluted by industrial pollutants.[8]

The Industrial Revolution made it possible for people to make goods more quickly and cheaply. It also made it easier for people to dispose of and replace the things they already had. Today, cheap clothing is easily made, but it is also easily thrown away. Many clothing companies focus on making inexpensive garments that follow recent trends. These items are often not durable and wear out quickly.

The business model that promotes buying and throwing out cheap, trendy clothing is called fast fashion. It contributes to land pollution in the form of wasted or unused clothes. Many people try to donate or resell old clothes. But there is far more secondhand clothing than there are people willing to buy it, which is why places such as the Atacama Desert become polluted with discarded clothes.

About 10 percent of greenhouse gas emissions in the United States in 2021 came from the agricultural industry, according to the EPA.[11]

In addition, clothes that are purchased and returned are often never resold because it is expensive for companies to repackage them. In the United States, about 66 percent of unwanted clothing is thrown away in landfills.[9] Experts estimate that close to six billion pounds (2.7 billion kg) of returned clothes end up in landfills every year.[10]

AGRICULTURAL POLLUTION

The agricultural industry releases pollutants into the land and water through the use of pesticides and fertilizers. In addition, agriculture is a major contributor to greenhouse gas emissions, especially methane. Scientists, activists, and farmers are searching for ways to grow and produce food without polluting the planet.

While some pesticides, such as DDT, have been banned, it is still common for farmers to use pesticides on their crops. These pesticides are used to kill insects, such as beetles and aphids, that are considered pests because they feed on the crops. Pesticides also kill plants that are considered weeds.

However, the chemicals in pesticides have far-ranging and long-lasting effects. Studies show that common pesticides are harmful to a majority of invertebrates that live in soil. These invertebrates, such as earthworms, play an important role in soil health and carbon storage. They help break down decomposing matter and distribute nutrients.

Several studies have found that pesticides harm or kill a majority of invertebrates living in soil. The chemicals from the pesticides also seep into underground water and can remain in

groundwater for decades. Some modern crop-growing materials and techniques can add other harmful substances to the soil as well. For example, some US strawberry growers use plastic mulch to prevent weeds and bacteria from growing. However, bits of plastic from the mulch can contaminate the soil, degrading the soil's quality and making it difficult for plants to grow.

Some agricultural pollutants come from livestock. Cows are large animals that require a lot of energy. The average dairy cow eats about 50 pounds (23 kg) of dry matter a day.[12] Dry matter is feed without water in it. A cow producing milk needs 30 to 50 gallons (114–189 L) of water a day.[13] Growing, harvesting, and transporting cow feed takes a great deal of energy, often involving the use of equipment such as tractors and trucks. These parts of the process release a significant amount of greenhouse gas emissions into the air.

In addition, when cows burp, they release methane. Methane is a greenhouse gas composed of hydrogen and carbon. It traps heat much more effectively than carbon dioxide. Livestock animals produce about 40 percent of current methane emissions, and most of those emissions come from cows.[14]

For these reasons, the beef industry is the biggest agricultural contributor to greenhouse gas emissions. In 2018, scientists found that producing about two pounds (1 kg) of beef released almost 220 pounds (100 kg) of greenhouse gases into the atmosphere.[15] Other livestock industries that have significant emissions include the dairy industry and the production of lamb.

SOLUTIONS TO INDUSTRIAL AND AGRICULTURAL POLLUTION

Some companies, organizations, and individuals are working to tackle the problems of industrial pollution. In the fashion industry, many companies are seeking ways to reduce the amount of wastewater they produce. For example, a typical pair of denim jeans takes about 2,866 gallons (10,850 L) of water to produce. This is partly because washing denim affects the color and look of the jeans.

Some companies, such as Outland Denim, have started using a process that uses lasers to get different jean looks. This process can reduce water usage by as much as 65 percent.[16]

Others are looking to reduce plastic use by shipping their products in recycled or nonplastic containers.

Many companies are also researching ways to create alternatives to fabrics such as rayon, a plant-based fabric that undergoes a water-intensive chemical process. One such new material is called lyocell. Like rayon, lyocell is made from plants, specifically wood. However, it is chemically treated with a closed-loop process, meaning that most of the water from the chemical treatment process is recycled instead of being released into the surrounding environment.

Several individuals and businesses in Kenya came together to hold a sustainable fashion event called Eco Fashion Week Kenya in November 2023.

Farmers who want to prevent agricultural pollution are finding ways to create sustainable farms. One approach is organic farming. Organic farming seeks to reduce the use of synthetic pesticides and other land pollutants. In order to be certified organic, a crop must be grown without using synthetic pesticides or fertilizers. Instead, organic farmers mostly rely on naturally derived pesticides, such as fungi, that are harmful to certain crop pests. According to the US government, the number of certified organic farms increased by more than 90 percent between 2011 and 2021.[18]

Water treatment plants use methods such as filtration and disinfection to ensure public drinking water is safe to consume.

WASTE, SANITATION, AND POLLUTION

Human waste is an environmental pollutant. Large human populations create huge amounts of waste from biological processes such as defecation. This waste can pollute water and damage ecosystems.

In addition, human garbage accumulates on land, in landfills, and in bodies of water. Wastewater processing and garbage management solutions help address the problems of waste and garbage. However, people are looking for more sustainable ways to process human waste and reduce trash.

DISEASES AND ILLNESSES

Humans produce pollutants with their bodies. Many common illnesses are caused by bacteria and viruses that spread from person to person. Body fluids such as blood, saliva, and mucus are very effective disease carriers. Human waste such as urine and feces is also a pollutant. Urine can spread a few types of diseases, including typhoid fever. But feces is more dangerous. It is full of microorganisms.

Disease-spreading bacteria and viruses also make their way into the digestive tract and human feces. If a person drinks water contaminated with feces, they can become sick. Worms and parasites are also spread through contact with feces. This process of infection and infestation is called the fecal-oral route of disease transmission. These infections often cause diarrhea.

Prolonged diarrhea can lead to severe dehydration, organ damage, and death. According to the World Health Organization (WHO), more than 800,000 people die every year from diarrhea caused by contaminated water.[1] Young children are especially vulnerable to illness from fecal-oral transmission.

In sub-Saharan Africa, many people walk more than 30 minutes to collect water. Oftentimes, that water is polluted and unsafe to drink.

SEWAGE

Human waste matter, both solid and liquid, is called sewage. Sewage can be human waste only, or it can be combined with other types of waste, such as industrial runoff or industrial sludge. The increase in the world's population has made sewage a significant pollution problem. On a small scale, natural environments are able to process the waste from animals, including humans. Bacteria in bodies of water break down fecal matter. And if there are larger, cleaner bodies of water available, the presence of small amounts of feces is less of a risk. However, in densely populated environments, the presence of sewage quickly becomes dangerous.

Human waste is hazardous to plant and animal life. Improperly treated waste released into nature can set off a chain reaction that disturbs the ecosystem. Many organisms, such as bacteria and algae, naturally feed on animal waste. Plants use animal waste as fertilizer.

When there is extra waste in a habitat, some organisms may thrive and overgrow. By doing so, they crowd out other forms of life that depend on the environment. They can also decrease the amount of oxygen available to plants and animals. Algae overgrowths in water are known as blooms. When these blooms overtake an environment, they can lead to the deaths of other species by producing toxic chemicals or blocking sunlight from reaching underwater plants. The blooms can also contaminate drinking water with the toxins they produce.

SANITATION

The process of managing sewage so that it does not spread disease and pollute the environment is called sanitation. About 88 percent of people globally use a sanitation service such as a sewer system with toilets. However, only about 57 percent of those systems are safely managed, according to the WHO.[2] People who lack access to toilets and sanitation must practice open defecation, where they eliminate waste in gutters or on land. This puts sewage into direct contact with local water and is a major health risk. And poorly managed sanitation systems can cause that sewage to remain in the environment.

Many countries have widespread access to flush toilet systems. A flush toilet disposes of sewage down a series of pipes so that it doesn't remain in the home. But in poorly managed sewer systems, flushed sewage may not be cleaned or treated before it is released back into water. In fact, 80 percent of the world's sewage is dumped into the ocean untreated.[3]

Since humans do not drink seawater, raw sewage in oceans is less of a direct threat to people's health. But sewage can still spread disease and disrupt marine ecosystems. Bacteria in oceans feed on raw human sewage. The bacteria populations grow and use up the oxygen in an area of the ocean, which leaves less oxygen for other life-forms to survive or grow. These areas are known as dead zones. Dead zones greatly threaten bays, lakes, and ocean coasts because they tend to receive more nutrients from upstream water sources such as rivers.

According to the UN, 42 percent of wastewater that comes from homes is not properly treated.[4]

Effective sanitation involves cleaning sewage from water before the water is released into reservoirs or natural bodies

of water. Wastewater plants do this by taking water from the sewer system and passing it through screens to remove solids. Then the water is exposed to oxygen and bacteria. The bacteria, with help from the oxygen, break down organic water pollutants. Finally, wastewater is treated with chlorine. The chlorine kills remaining organisms not caught by earlier stages of the process.

Some treatment plants may take extra steps to remove things such as phosphates, which are commonly found in fertilizers. This process makes wastewater fit for human consumption. The treated water may then find its way back into taps.

Most water in the taps of US homes comes from places such as rivers, streams, and reservoirs. After plants treat this water to make it safe for human consumption and for organisms to live in, they release it back into the environment. However, the wastewater treatment process is often imperfect.

For example, many chemicals from cleaning products and detergents are flushed down drains and end up in wastewater treatment plants. Many wastewater plants are not equipped to process these chemicals, so the chemicals end up in bodies of water. Wastewater plants are also often not effective at filtering drugs and hormones from the water supply. Chemicals from human and animal pharmaceuticals enter water through urine and feces. These chemicals can stay active even after water is treated.

> **We can't have landfills forever, and we can't ask others to accept our trash.**[5]
>
> *—Jaime Lerner, former Brazilian politician, 2016*

LANDFILLS

Garbage dumps are a very old method of getting rid of trash. Decomposing garbage produces greenhouse gases and unpleasant smells. It also attracts pests. Therefore, it makes sense to store garbage away from human homes. Some waste could be fed to animals, such as pigs, or collected and reused. However, the amount of garbage people produce has increased over time, especially in developed countries. In 1960, the average American produced 2.7 pounds (1.2 kg) of solid trash every day. By 2018, that number had increased to 4.9 pounds (2.2 kg).[6]

Landfills were invented to help store the large quantities of human trash. These areas are specially set aside for solid garbage. Landfills are often built by digging holes in the ground with the intention of filling the empty space with garbage as time goes on. The first landfills were simply areas where holes were dug into the ground. Today, landfills are prepared with liners that separate garbage from the dirt. This is to prevent chemicals from seeping into the ground. Landfills are divided into sections, also known as cells. In each cell, trash is layered and then compacted so the trash inside the cell takes up less space.

In the United States, more than half of solid garbage ends up in landfills. Landfills can cause a variety of pollution issues.

First, they make land unusable for wildlife. When they are open and accepting trash, landfills are unsafe for humans and animals, and there is no soil available for plants to grow. After a landfill is closed, it is capped with a layer of soil. However, the soil above closed landfills tends to show signs of exposure to garbage. The dirt on top of closed landfills has more harmful heavy metals in it than other types of soil. This is because materials from decomposing garbage make their way into the soil.

In addition, landfills release large amounts of methane and carbon dioxide, as well as other greenhouse gases. This is because the garbage inside a landfill does not decompose the same way it would in nature. When organic material such as old food breaks down in nature, it is exposed to oxygen in the air. Trash that is compacted and buried underground has little access to oxygen. It breaks down differently and more slowly. The process of trash decomposition produces greenhouse gases, especially carbon dioxide and methane. Some landfills can emit greenhouse gases for more than 50 years after being constructed.[8]

In 2022, the world's largest landfill was in Sudokwon, South Korea. It took in 20,000 short tons (18,150 metric tons) of waste every day.[9]

Landfills are also potential sources of water pollution. When landfill liners leak, runoff from garbage can pollute groundwater. Some landfills are safer and better constructed than others. But in many cases, leaks are impossible to prevent completely.

Exposure to landfills and their harmful effects is unevenly distributed throughout the world. Many developed countries, including the United States and Canada, ship their garbage to other countries. Moving trash around the world is a profitable business, but it releases significant greenhouse gas emissions. For example, in 2021, Malaysia imported more than

Wild animals are often drawn to landfills because of the food scraps people throw away.

Mixed recycling facilities can sort materials by hand or by using machines.

500,000 short tons (454,000 metric tons) of plastic garbage from other countries.[10] This waste is often intended for recycling. But more often, it cannot be safely or efficiently reused.

RECYCLING, GARBAGE MANAGEMENT, AND WASTEWATER SOLUTIONS

Because sanitation is key to human health, ensuring safe management of water and waste is a priority for the international community. The United States dedicates some of its foreign aid money to building sanitation services and making drinking water safe. Between 2008 and 2022, the United States helped more than 50 million people get access to sanitation services.[11]

The UN aims to achieve clean water and sanitation for all people by 2030. To do this, it is investing in new water-cleaning technology. It also calls on national governments to make sanitation a high priority.

Many people around the world are looking for solutions for managing waste. One of the most important tools to reduce reliance on landfills is recycling, which is when a used material is reprocessed so that it can be used again. This reduces the amount of waste that needs to be burned or put in landfills. Some products are more easily recycled than others.

For example, aluminum is a metal often used to make cans and auto parts. It is easy to clean and reshape, making it good for recycling. More than 75 percent of the aluminum ever produced is still being used today.[12] Some companies are even switching from plastic to aluminum packaging as a way to increase the recycling efficiency of their products. The bottled water company Open Water sells all its products in aluminum cans rather than plastic bottles.

Many governments are also looking for ways to reduce the amount of trash that goes into landfills. New York City's government aims to eliminate landfill trash by 2030. To reach this goal, the city requires all businesses to recycle. In addition, it requires some large businesses to sort garbage for composting.

Composting is a method for turning organic garbage, or trash made from once-living things, into a dirt-like material. In a compost heap or container, microorganisms, water, and oxygen break down this garbage. The resulting product can be used as a fertilizer for soil.

Trash Pickers and Recycling by Hand

In some countries, people make a living from sorting, recycling, and reusing garbage by hand. In Cairo, Egypt, a group of people called the Zabbaleen survive by collecting the city's waste, sorting it, and reselling it for a profit. In the past, the group also kept pigs that would eat food scraps sourced from city garbage. People in other countries do similar jobs. In Indonesia, trash pickers brave hazardous landfills to find recyclable plastics that they can resell. This is often dangerous work because of the heavy machinery being used in the landfills and the potential for a landslide to occur due to shifting garbage.

AUTUMN PELTIER

Autumn Peltier is a Canadian activist who fights for many causes, including clean water access. Peltier is a member of the Anishinabek Nation, a group of people native to the northern United States and Canada. She is the chief water commissioner of the Anishinabek Nation. When Peltier was a child, she visited an Indigenous community that did not have access to clean tap water. Peltier's experience encouraged her to speak in favor of government policies that would ensure clean water for all.

In 2016, at age 12, Peltier met Canadian prime minister Justin Trudeau. Peltier protested the Canadian government's decision to install new pipelines to pump bitumen, an oil by-product. These pipelines, which crossed through Indigenous lands, have a high likelihood of leaking and releasing pollutants into the soil and waterways.

Since her meeting with Trudeau, Peltier has continued to speak out about preserving water. In 2018 and 2019, she gave speeches at UN conferences. Peltier believes that young people should have a greater voice in decisions about pollution and preserving natural resources. "Water justice means no more broken promises from [the] government," Peltier says.[13]

Autumn Peltier was named Canada's Walk of Fame Community Hero in 2023.

Plastic bottles, bags, and utensils are common items found in polluted bodies of water.

CHAPTER 7

WATER POLLUTION

Life on Earth depends on water. About 70 percent of the planet is covered in water.[1] Approximately 55 to 60 percent of the human body is water as well.[2] Human beings depend on water, specifically fresh water, to survive.

Fresh water is water that has less than 1 percent salt concentration. Less than 3 percent of Earth's water is fresh, and most of that water is stuck in glaciers.[3] Increasing water pollution is making clean, drinkable fresh water even scarcer. According to the Centers for Disease Control and Prevention, two billion people around the world do not have access to clean, safe drinking water.[4] Pollution causes complex and long-lasting damage to freshwater biomes such as lakes, rivers, and streams.

The salt water of the world's oceans is at risk too. Ocean transportation, trash dumping, sewage mismanagement, and chemical contamination have all worsened ocean pollution. In addition, climate change caused by greenhouse gas emission has led to changes in sea levels. Activists, scientists, and politicians are working together to preserve the oceans and clean water supplies around the world.

FRESHWATER POLLUTION

In 2021, a UN team studied 75,000 bodies of water in 89 countries. The team found that 40 percent of these bodies of water had a level of pollution described as severe.[5] In 2022, a nonprofit group called the Environmental Integrity Project did a similar study of rivers and streams. It found that about 700,000 miles (1.1 million km) of the US rivers and streams assessed were too polluted for people to fish and swim in or drink from.[6]

When freshwater sources are polluted, all local life suffers. It can be very difficult for a water source to recover once pollution has occurred. One example is the Yellow River in China. This 3,400-mile (5,500 km) river stretches from the interior of the country to the eastern coast.[7] It gets its name from the yellow color of the silt, or fine clay, the waters carry.

Throughout China's history, many people have settled in cities and towns along the banks of the Yellow River. For this reason, the river is sometimes called the cradle of Chinese civilization. Today, the Yellow River provides water to more than 50 Chinese cities. Its water is also used to irrigate 15 percent of China's farmland.[8]

The Yellow River is home to unique wildlife, such as the Yellow River catfish. Rare mammals, including a type of antelope known as the chiru, live along the river. Its marshes and the nearby land are also home to more than 300 species of birds.[10]

Many cities along the Yellow River are major manufacturing hubs. Lanzhou is a center for oil and mining industries and a producer of industrial chemicals and fertilizers. The city's economy grew rapidly in the 1900s. But factories would commonly dump untreated or barely treated sludge into the Yellow River.

In 2008, research showed that the river was in crisis. Water samples found that one-third of the river's water was too polluted to drink, swim in, or use for fishing or farming.[11] A study found that between 1965 and 2015, the number of fish species in the river decreased by more than a third.[12]

The Chinese government has tackled some of the factors that led to the severe pollution in the Yellow River. In 2022, the Yellow River Protection Law was passed. This law bans any new

Frog Populations

One of the unexpected effects of water pollutants is the way they have changed the biology of some water-dwelling animals. Scientists have found evidence that chemical pollutants, including the pesticide atrazine, can change the way frogs reproduce. Atrazine prevents weeds from growing amid crops such as corn. In the early 2000s, scientists found that male frogs exposed to atrazine often could not reproduce. Since then, other studies have confirmed that common pesticides in water change frog biology. These changes could lead to a decline in frog populations.

manufacturing of chemicals alongside the river. It also requires local governments to tighten their water pollution standards if the river's water falls below a certain quality level.

Pollution can also lead to the permanent loss of some kinds of wildlife. For example, the Yangtze River in China was once the home to a species of river dolphin called the baiji. Baiji could not see well and depended on echolocation to hunt and navigate. The Yangtze River has heavy ship traffic, and the noise

The baiji is also known as the Yangtze River dolphin.

pollution from boats tended to confuse the animals. Many baiji died from collisions with ships. In addition, water pollution made their habitats harder to live in. Toxic chemicals from industrial and agricultural processes polluted the environment and were absorbed into the animals' bodies. Baiji were also hunted for their meat and oil. All these factors eventually led to the species being declared extinct in late 2006. The baiji was the first whale or dolphin to go extinct in the modern age.

Garbage also contributes to freshwater pollution. And eventually, that garbage becomes the source of ocean pollution. Plastic and other trash flow from freshwater sources into the sea. Just a thousand of the rivers worldwide are the source of 80 percent of ocean plastic pollution.[13] This plastic can wreak its own havoc in marine waters. And since many rivers open into the ocean, chemicals from industrial and agricultural pollution also make their way into the sea.

In 2022, 50 percent of rivers and streams and 55 percent of ponds, lakes, and reservoirs in the United States were so polluted they weren't safe for human use.[15]

OCEAN POLLUTION

Today, 80 percent of the world's shipping happens through ocean waters.[14] If ships leak fuel or cargo, ocean life can be affected. When marine birds and marine mammals with fur are covered in oil, their feathers and fur can no longer keep them warm. The animals may die of hypothermia, or extreme cold. Ocean animals

Water Pollution and Wildlife

Pieces of plastic waste can attach to ocean wildlife, causing injury. The plastic rings that bundle six-packs of soda cans are usually made of plastic. When these rings are discarded in water, they can easily strangle smaller animals such as turtles. Plastic straws are similarly dangerous. They can block the airways and stomachs of animals, including seabirds and dolphins. Some companies are experimenting with plastic packaging alternatives. For example, one beverage company uses wheat and barley to make compostable six-pack rings.

can also die from breathing in or swallowing oil. Oil in ocean water can damage fish and coral as well, leading to infertility, illness, and death. Large oil spills, which are spills that release more than 700 short tons (635 metric tons) of oil, are less frequent today than they were in the 1990s. Still, during the 2010s, there was an average of almost two large oil spills each year.[16]

One of the biggest challenges the ocean faces is trash. The ocean has become a home to giant masses of garbage that mostly came from land. Wind, storm drains, and freshwater flows carry the garbage to the ocean. Some of it originates from the sea in objects such as fishing nets. It also sometimes floats together in giant patches. The biggest of these is known as the Great Pacific Garbage Patch.

The Great Pacific Garbage Patch consists of two smaller collections, a Western Patch and an Eastern Patch. These patches travel around the middle of the Pacific Ocean. The Western Patch is close to Japan, and the Eastern Patch is near Hawaii. Together, they cover an area twice the size of Texas.

Much of the garbage in the Great Pacific Garbage Patch is not visible to the naked eye. Instead, the garbage has broken down into microplastics, which are pieces of plastic less than 0.2 inches (0.5 cm) long. Microplastics travel throughout the ocean. In fact, most of the plastic in the ocean isn't on the surface at all. It has sunk into the deep ocean. The bottom of the ocean is home to 10,000 times more microplastic than the ocean surface.[17] Scientists still don't know all the effects of microplastics. But they do know they are now everywhere in the ocean. A 2023 study sampled water from 45 ocean locations around the globe and found microplastics in every sample.

CLIMATE CHANGE AND WATER POLLUTION

The ocean is a carbon sink. Like forests, oceans play an important role in pulling carbon from the atmosphere. Since carbon emissions have increased, the amount of carbon absorbed by the ocean has too. This rise in carbon absorption makes the ocean more acidic. Increased ocean acidity can have many harmful effects on wildlife. For example, higher acidity makes it more difficult for animals such as oysters to form shells.

In addition, because of climate change, sea levels are rising in many parts of the world. This means the ocean reaches parts of land that it previously did not. Rising sea levels brings its own risks of pollution, such as saltwater contamination of freshwater sources. Miami, Florida, is one place threatened by rising sea levels. The aquifer, or underground water source, that provides the city with water is exposed to more salt water as sea levels rise. Someday the water from the aquifer may become too salty for people to drink. This is a kind of natural pollution that climate change makes more likely to occur.

SOLUTIONS FOR WATER POLLUTION

Freshwater contamination is a problem that cannot be ignored. Like China, many countries are prioritizing river cleanups. In India, the Ganges River is a water source, the center of an ecosystem, and a cultural and religious site. Many people practicing Hinduism believe that swimming or bathing in the Ganges can bring healing. However, the Ganges is heavily polluted by industrial runoff and sewage.

Since 2014, the Indian government has committed to cleaning the Ganges. In 2022, the country had completed

In 2023, the Ocean Cleanup tested its new Interceptor 006 Barricade to capture plastic waste in the Rio Las Vacas in Guatemala.

177 of its 363 projects designed to help clean the river and restore biodiversity.[19] One major concern is the amount of plastic that ends up in the river. In 2022, India banned all single-use plastics, including plastic bags and utensils. India has also planted trees along the banks of the Ganges. These trees can act as barriers to keep pollutants out of the water.

Other organizations are tackling the problem of ocean plastics. The Ocean Cleanup is a nonprofit organization dedicated to fighting plastic pollution. It creates technology to capture and process plastic waste in bodies of water. One of its inventions is

a boat-like machine called the Interceptor. This machine sits in a river, takes in water, and removes plastic before the waste reaches the ocean. The Ocean Cleanup also places large, slow-moving nets near the surface of locations with lots of plastic pollution in the water. These floating systems gather up plastics to make extracting them easier.

Buying secondhand clothing reduces natural resource use and greenhouse gas emissions.

TAKING ACTION AGAINST POLLUTION

There are many ways for people to take action in the fight against pollution. People can make everyday choices that reduce the impact of pollution. They can join or create research projects that help address key pollution challenges. Or they can partner with local and global organizations that work to clean the planet and advocate for widespread changes.

INDIVIDUAL STEPS TO FIGHT POLLUTION

Some activists find inspiration in making small, personal steps to fight pollution. For example, buying secondhand clothing, swapping clothes with friends and relatives, and mending clothes by hand are ways to reduce the waste produced by the fashion industry. Others may support clothing brands that recycle materials or try to conserve water in the manufacturing process.

Some people even take things a step further and make their own clothes out of their own materials. In Nigeria, a group of teens combine fashion innovation with environmental action. This group organizes a yearly "Trashion Show." They make outfits and accessories from plastic waste and display them in a runway show. The show helps raise money to help clear a local waterway blocked by garbage. The teens are passionate about fighting environmental pollution. "We need to start now because in a few years it's going to be too late to do anything," said group member Esohe Ozigbo in 2021.[1]

When it comes to agricultural pollution, people can learn about better ways to grow food. For example, Greenagers is an organization in the northeast United States that connects teens to jobs and internships in farming and conservation. Those who join Greenagers learn farming practices that can help reduce pollution and restore the health of local soil.

4-H Clubs

4-H is a network of youth organizations closely linked to agriculture. Students in 4-H clubs learn not only about farming but also science, citizenship, healthy living, and more. Some youth in 4-H have become involved in solutions to soil pollution. In Illinois and Ohio, 4-H programs teach members how to analyze the contents of soil. And in Minnesota, 4-H members learn about biodegradable plastics.

INNOVATIONS IN THE FIGHT AGAINST POLLUTION

Some activists are using their science expertise to innovate new pollution solutions. In 2016, four teens joined a research project investigating whether certain kinds of worms could help break down plastic waste. Working with scientists, the teens found that mealworms will eat plastic. They also found that the worms seem to prefer certain kinds of plastic to others.

Teen Joseph Galasso, who took part in the project, also did his own research on plastic-eating bacteria. He analyzed the gut bacteria of superworms, a larger mealworm relative. Galasso used a machine to analyze the DNA of the bacteria inside the worms and was able to prove that like mealworms, superworms used the bacteria to digest plastic. Galasso hopes his research can help people find ways to use the worms as a means of reducing plastic pollution.

Another way people are taking action is by inventing new technology. In 2018, when she was just 12, American Anna Du invented a robot that could sense microplastics on the ocean floor. Her robot could help pave the way to future ocean cleanups. During that time, Du also created the Deep Plastics Initiative, a nonprofit aimed at cleaning up plastic pollution, especially in the ocean. The nonprofit brings together researchers, engineers, and educators who are interested in creating new technology to fight pollution.

VOLUNTEER EFFORTS

One way to counteract air pollution from carbon emissions is to plant trees. Through photosynthesis, trees absorb carbon from the air and release oxygen into the atmosphere. The bigger and older the tree, the more carbon it can absorb. Some people are taking the lead in tree-planting initiatives.

When he was in fourth grade, German environmentalist Felix Finkbeiner planted his first tree. By his teens, his passion for tree planting led him on a mission to plant one tree for everyone in the world. Finkbeiner has helped in the planting of more than ten billion trees. He founded an organization called Plant for the Planet. Plant for the Planet helps young people get involved in tree planting. The goal of the organization is to plant one trillion trees.

> You have power as an individual. It can feel tiny, but your actions matter.[3]
>
> —Melati Wijsen, environmental activist

ORGANIZATIONS THAT FIGHT POLLUTION

In the United States, many organizations have ways for people to work toward clean water, air, and soil. SustainUs and the Sunrise Movement are both youth-led organizations engaged in the fight against pollution. SustainUs encourages teens to help reduce the effects of pollution and climate change. The organization sends groups of young people to conferences such as the World Bank Annual Meeting to speak in favor of eliminating fossil fuel use.

The organization believes that young people are especially affected by climate change and pollution, and their voices need to be represented.

The Sunrise Movement advocates for the Green New Deal, a proposed government policy that would push the United States to use more renewable energy. The organization creates campaigns to bring attention to pollution-related policies. It also works to elect more officials that support pollution solutions.

People who get involved in antipollution activism find inspiration in knowing their efforts make a difference. In 2016, sisters Amy and Ella Meek learned about the problem of ocean plastic and decided to start their own organization, Kids Against Plastic. The sisters have developed a plan to help schools use plastic more responsibly. Amy said, "Think of something you are passionate about, and then break it down into small, manageable chunks that will enable you to start doing something—no matter how small."[4]

In April 2023, several US lawmakers held a press conference to reintroduce the Green New Deal in hopes of enacting the policy.

Today, pollution touches every part of life. It's in the air, water, and land. The food people eat every day, the cars they depend on, the heat that makes their homes livable, and the clothes they wear can all contribute to pollution. In addition, pollution can affect any ecosystem and the organisms that live there.

While some of the effects of pollution are reversible, it can also lead to irreversible damage, such as the extinction of a species. For these reasons, pollution can feel like a vast and overwhelming problem. But many people, businesses, and governments around the globe have not given up hope for a cleaner planet and future. Through everyday acts, public advocacy, and research, people are proving the importance of taking action to combat the effects of pollution and create a more sustainable way of life.

Speak Up

One of the most important ways to combat pollution in any form is to speak up. People can speak up in simple ways, such as talking to friends and family about issues related to pollution. Educating others about pollution can help spread the word and motivate them to find solutions. People can also speak up in bigger ways. Directly contacting businesses that are contributing to pollution, especially plastic pollution, is a great way to push those businesses to use more sustainable products and practices. Contacting local, state, or federal government officials can help get more laws put into effect that will reduce pollution going into the environment.

POLLUTION PROBLEMS

- Air pollution from fossil fuel emissions, such as car and factory exhaust, warms the planet and threatens ecosystems and wildlife.

- Industrial and agricultural pollution can release harmful chemicals and greenhouse gases, contaminating the air, land, and water.

- Human waste that isn't treated properly can end up in water and soil, disrupting ecosystems and causing illnesses and diseases.

- Trash and plastics often end up in landfills, producing large amounts of greenhouses gases and contributing to climate change.

- Rivers, lakes, oceans, and other bodies of water face pollution from sewage, industrial waste, littering, and disasters such as oil spills.

POLLUTION SOLUTIONS

- Decreasing the use of fossil fuels, such as coal and oil, to reduce greenhouse gas emissions.

- Using sustainable business practices to save water and prevent pollution from synthetic chemicals.

- Practicing sustainable farming techniques to help reduce the use of synthetic pesticides and lower methane emissions from farm animals.

- Treating wastewater effectively to protect humans, animals, and habitats.

- Recycling and reducing plastic consumption to limit the amount of trash and plastic that ends up in waterways and landfills.

- Contact government officials to get more laws and regulations put in place to fight pollution.

- Volunteer to clean up garbage in waterways or parks, or volunteer to plant trees.

- Cut down on plastic consumption by using reusable products, such as bags and straws.

QUOTE

"If we are going to live so intimately with chemicals—eating and drinking them, taking them into the very marrow of our bones— we had better know something about their power."

—*Rachel Carson*, Silent Spring, *1962*

alga
A plantlike organism that grows mostly in water and is often green, blue, red, or brown colored.

biodiversity
The many different plants and animals in an ecosystem.

brackish
Slightly salty, such as the conditions in the water where river and seawater mix in estuaries.

compound
A substance made of two or more elements.

decompose
To decay and become rotten.

DNA
Deoxyribonucleic acid, the chemical that is the basis of genetics, through which various traits are passed from parent to child.

echolocation
A process of using sound waves to locate objects in an area.

ecosystem
A community of interacting organisms and their environment.

emission
The production and discharge of something such as smoke, gas, or chemicals into the air.

invertebrate

An animal without a spinal column.

irrigation

A system that brings water from one location to an area that has crops.

metallurgy

The process of extracting metal from ore so the metal can be used.

pharmaceutical

A medicinal drug.

ratify

To formally approve or adopt an idea or document.

synthetic

A material made through artificial chemical processes rather than by nature.

toxic

Poisonous to an animal or habitat.

turbine

A machine that uses a fast-moving flow of water, steam, gas, air, or other fluid to produce energy.

SELECTED BIBLIOGRAPHY

Lindwall, Courtney. "Industrial Agricultural Pollution 101."
Natural Resources Defense Fund, 21 July 2022, nrdc.org. Accessed
11 Dec. 2023.

McCracken-Heywood, Neve. "Is Sewage Pollution Still a Major
Threat to Our Oceans?" *Big Blue Ocean Cleanup*, 4 Nov. 2021,
bigblueoceancleanup.org. Accessed 11 Dec. 2023.

"Sources and Solutions: Fossil Fuels." *Environmental Protection Agency*,
29 Nov. 2023, epa.gov. Accessed 11 Dec. 2023.

FURTHER READINGS

Andrus, Aubre. *The Plastic Problem: 60 Small Ways to Reduce Waste and
Help Save the Earth*. Lonely Planet, 2020.

Clendenan, Megan. *Fresh Air, Clean Water: Our Right to a Healthy
Environment*. Orca Book Publishers, 2022.

Mooney, Carla. *Climate Change*. Abdo, 2025.

ONLINE RESOURCES

To learn more about pollution, please visit
abdobooklinks.com or scan this QR code.
These links are routinely monitored and
updated to provide the most current
information available.

MORE INFORMATION

For more information on this subject, contact or visit the following organizations:

Beat Pollution

2 UN Plaza, Room DC2-803

323 E. 44th St.

New York, NY 10017

unep.org

Beat Pollution is a program under the United Nations Environment Programme working toward a pollution-free planet. The program aims to eliminate all forms of pollution to better the health and safety of people, animals, and environments around the world.

WaterAid

233 Broadway, Ste. 2705

New York, NY 10279

wateraid.org

WaterAid is a global organization working to improve access to clean water in lower-income communities. The organization's projects include educating people about good hygienic practices and providing communities with toilets.

World Health Organization (WHO)

525 23rd St. NW

Washington, DC 20037

who.int

The World Health Organization (WHO) is an organization that works to provide better global health care and improve people's health. The WHO investigates the factors that contribute to certain health problems, such as pollution, and looks for solutions to create a healthier world for everyone.

CHAPTER 1. SAVING YOSEMITE

1. "Yosemite Facts: Yosemite National Park." *Yosemite Mariposa County*, n.d., yosemite.com. Accessed 1 Feb. 2024.

2. Tik Root. "Cigarette Butts Are Toxic Plastic Pollution. Should They Be Banned?" *National Geographic*, 9 Aug. 2019, nationalgeographic.com. Accessed 2 Feb. 2024.

3. "Central Valley." *Britannica*, 29 Jan. 2024, britannica.com. Accessed 1 Feb. 2024.

4. "How the Central Valley Feeds the Nation." *Fruit Growers Supply Company*, 19 Jan. 2023, fruitgrowers.com. Accessed 1 Feb. 2024.

5. Yi Hyun Kim, Shen Xin, and Jackson Fordham. "In California's Heartland, Small-Time Almond Farmers Face a Dry Future." *Points Unknown*, 2022, pointsunknown.nyc. Accessed 1 Feb. 2024.

CHAPTER 2. THE SCIENCE OF POLLUTION

1. "Noise Pollution." *National Geographic*, 19 Oct. 2023, education.nationalgeographic.org. Accessed 1 Feb. 2024.

2. "Noise Pollution."

3. Jerry A. Nathanson. "Light Pollution." *Britannica*, 31 Jan. 2024, britannica.com. Accessed 1 Feb. 2024.

4. Robert W. Decker, Barbara B. Decker, and Sanat Pai Raikar. "Volcanic Eruption." *Britannica*, 16 Jan. 2024, britannica.com. Accessed 1 Feb. 2024.

5. "The Lifecycle of Plastics." *World Wildlife Fund Australia*, 1 July 2021, wwf.org.au. Accessed 1 Feb. 2024.

6. Leonardo DiCaprio (@leonardodicaprio). 2020. "Plastic pollution is a global issue: killing wildlife, contaminating our oceans and waterways, and lasting far longer than it is used." *Instagram*, 21 Aug. 2020, instagram.com. Accessed 1 Feb. 2024.

7. Katie Strick. "'Greta Thunberg and I Are Living Proof That Kids Can Change the World for Good.'" *Standard*, 23 Feb. 2023, standard.co.uk. Accessed 1 Feb. 2024.

8. "Overview of EPA's Brownfields Program." *Environmental Protection Agency*, n.d., epa.gov. Accessed 1 Feb. 2024.

9. Marlene Lenthang and Phil Helsel. "Air Quality Concerns Continue as Canadian Wildfire Smoke Covers the Northeast." *NBC News*, 7 June 2023, nbcnews.com. Accessed 1 Feb. 2024.

10. Hannah Ritchie, Fiona Spooner, and Max Roser. "Clean Water." *Our World in Data*, Sept. 2019, ourworldindata.org. Accessed 1 Feb. 2024.

11. Hannah Ritchie and Max Roser. "Air Pollution." *Our World in Data*, Oct. 2017, ourworldindata.org. Accessed 1 Feb. 2024.

12. Damian Carrington. "Microplastics Found in Human Blood for First Time." *Guardian*, 24 Mar. 2022, theguardian.com. Accessed 1 Feb. 2024.

CHAPTER 3. THE HISTORY OF POLLUTION

1. "Agent Orange." *Britannica*, 23 Jan. 2024, britannica.com. Accessed 1 Feb. 2024.

2. "Population Change." *Britannica*, n.d., britannica.com. Accessed 1 Feb. 2024.

3. Ian Tiseo. "Cumulative Carbon Dioxide (CO2) Emissions from Fossil Fuel Combustion Worldwide from 1750 to 2022, by Major Industry." *Statista*, 12 Dec. 2023, statista.com. Accessed 1 Feb. 2024.

4. Hannah Ritchie and Max Roser. "United States: CO2 Country Profile." *Our World in Data*, 2023, ourworldindata.org. Accessed 1 Feb. 2024.

5. Rachel Carson. *Silent Spring*. Houghton Mifflin, 1962, p. 17.

6. "*Exxon Valdez*." *National Oceanic and Atmospheric Administration*, 17 Aug. 2020, darrp.noaa.gov. Accessed 1 Feb. 2024.

CHAPTER 4. FOSSIL FUELS AND POLLUTION

1. "OPEC Share of World Crude Oil Reserves, 2022." *Organization of the Petroleum Exporting Countries*, 2023, opec.org. Accessed 1 Feb. 2024.

2. Andrew Fawthrop. "Profiling the Top Five Countries with the Biggest Natural Gas Reserves." *NS Energy*, 15 Mar. 2021, nsenergybusiness.com. Accessed 1 Feb. 2024.

3. "What Is US Electricity Generation by Energy Source?" *United States Energy Information Administration*, Oct. 2023, eia.gov. Accessed 1 Feb. 2024.

4. "Coal." *International Energy Agency*, 11 July 2023, iea.org. Accessed 1 Feb. 2024.

5. "Years of Fossil Fuel Reserves Left, 2020." *Our World in Data*, 2023, ourworldindata.org. Accessed 1 Feb. 2024.

6. "Fossil Fuels." *National Geographic*, 19 Oct. 2023, education.nationalgeographic.org. Accessed 1 Feb. 2024.

7. Alec Tyson, Cary Funk, and Brian Kennedy. "Americans Largely Favor US Taking Steps to Become Carbon Neutral by 2050." *Pew Research Center*, 1 Mar. 2022, pewresearch.org. Accessed 1 Feb. 2024.

8. "What to Know before Purchasing an Electric Vehicle: A Buying Guide." *Cars.com*, 28 Mar. 2023, cars.com. Accessed 1 Feb. 2024.

9. António Guterres. "Secretary General's Press Conference on Climate." *United Nations*, 27 July 2023, un.org. Accessed 1 Feb. 2024.

10. "Emissions from Electric Vehicles." *United States Department of Energy*, n.d., afdc.energy.gov. Accessed 1 Feb. 2024.

11. "Wind Explained: Electricity Generation from Wind." *United States Energy Information Administration*, Feb. 2023, eia.gov. Accessed 1 Feb. 2024.

12. "Emissions from Electric Vehicles."

13. "Marking the Kyoto Protocol's 25th Anniversary." *United Nations*, 2022, un.org. Accessed 1 Feb. 2024.

CHAPTER 5. INDUSTRIAL AND AGRICULTURAL POLLUTION

1. "Sources of Greenhouse Gas Emissions." *Environmental Protection Agency*, n.d., epa.gov. Accessed 1 Feb. 2024.

2. "Fashion's Problems." *Clean Clothes Campaign*, n.d., cleanclothes.org. Accessed 1 Feb. 2024.

3. "Top 10 Manufacturing Countries in 2023." *Safeguard Global*, 15 Sept. 2023, safeguardglobal.com. Accessed 1 Feb. 2024.

4. Tiffany Duong. "Chile's Atacama Desert: Where Fast Fashion Goes to Die." *EcoWatch*, 12 Nov. 2021, ecowatch.com. Accessed 1 Feb. 2024.

5. "Environmental Sustainability in the Fashion Industry." *Geneva Environment Network*, 29 Nov. 2023, genevaenvironmentnetwork.org. Accessed 1 Feb. 2024.

6. Taylor Mogavero. "Clothed in Conservation: Fashion & Water." *Florida State University*, 16 Apr. 2020, sustainablecampus.fsu.edu. Accessed 1 Feb. 2024.

7. Mogavero, "Clothed in Conservation."

8. Kathleen Webber. "How Fast Fashion Is Killing Rivers Worldwide." *EcoWatch*, 22 Mar. 2017, ecowatch.com. Accessed 1 Feb. 2024.

9. Arabella Ruiz. "17 Most Worrying Textile Waste Statistics & Facts." *TheRoundup.org*, 11 Apr. 2023, theroundup.org. Accessed 1 Feb. 2024.

10. Alina Selyukh. "From Living Rooms to Landfills, Some Holiday Shopping Returns Take a 'Very Sad Path.'" *National Public Radio*, 12 Jan. 2022, npr.org. Accessed 1 Feb. 2024.

11. "Sources of Greenhouse Gas Emissions."

12. "How Many Pounds of Feed Does a Cow Eat in a Day?" *DAIReXNET*, 16 Aug. 2019, dairy-cattle.extension.org. Accessed 1 Feb. 2024.

13. Craig Thomas. "Drinking Water for Dairy Cattle: Part 1." *Michigan State University*, 5 Apr. 2011, canr.msu.edu. Accessed 1 Feb. 2024.

14. Christopher Booker and Sam Weber. "Cow Burps Are a Major Contributor to Climate Change—Can Scientists Change That?" *PBS*, 6 Mar. 2022, pbs.org. Accessed 1 Feb. 2024.

15. Hannah Ritchie, Pablo Rosado, and Max Roser. "Environmental Impacts of Food Production." *Our World in Data*, 2022, ourworldindata.org. Accessed 1 Feb. 2024.

16. Emily Chan. "The Fashion Industry Is Using Up Too Much Water—Here's How You Can Reduce Your H2O Footprint." *Vogue India*, 22 Mar. 2020, vogue.in. Accessed 1 Feb. 2024.

17. Walton Family Foundation. "Majority of Americans Want Congress to Reauthorize the Farm Bill, in Favor of Promoting Sustainable Agricultural Practices to Protect Water, Reduce Air Pollution." *PR Newswire*, 18 Jan. 2023, prnewswire.com. Accessed 1 Feb. 2024.

18. "Organic Agriculture: Overview." *United States Department of Agriculture*, n.d., ers.usda.gov. Accessed 1 Feb. 2024.

CHAPTER 6. WASTE, SANITATION, AND POLLUTION

1. Esteban Ortiz-Prado, et al. "Waterborne Diseases and Ethnic-Related Disparities: A 10 Years Nationwide Mortality and Burden of Disease Analysis from Ecuador." *National Library of Medicine*, 21 Dec. 2022, ncbi.nlm.nih.gov. Accessed 1 Feb. 2024.

2. "Sanitation." *World Health Organization*, 3 Oct. 2023, who.int. Accessed 1 Feb. 2024.

3. Neve McCracken-Heywood. "Is Sewage Pollution Still a Major Threat to Our Oceans?" *Big Blue Ocean Cleanup*, 4 Nov. 2021, bigblueoceancleanup.org. Accessed 1 Feb. 2024.

4. "Water Quality and Wastewater." *United Nations Water*, n.d., unwater.org. Accessed 1 Feb. 2024.

5. David Adler. "Story of Cities #37: How Radical Ideas Turned Curitiba into Brazil's 'Green Capital.'" *Guardian*, 6 May 2016, theguardian.com. Accessed 1 Feb. 2024.

6. "National Overview: Facts, Figures on Materials, Wastes, and Recycling." *Environmental Protection Agency*, n.d., epa.gov. Accessed 1 Feb. 2024.

7. Sarah Golden. "What's Wrong with Burning Garbage?" *GreenBiz*, 21 Aug. 2020, greenbiz.com. Accessed 1 Feb. 2024.

8. "Important Things to Know about Landfill Gas." *New York State Department of Health*, n.d., health.ny.gov. Accessed 1 Feb. 2024.

9. Ian Tiseo. "Daily Volume of Waste Dumped at the Largest Landfills Worldwide as of 2022." *Statista*, 6 Feb. 2023, statista.com. Accessed 1 Feb. 2024.

10. "Trade Volume of Plastic Waste in Malaysia in 2021." *Statista*, 27 Jan. 2023, statista.com. Accessed 1 Feb. 2024.

11. "Water and Sanitation." *United States Agency for International Development*, n.d., usaid.gov. Accessed 1 Feb. 2024.

12. "Infinitely Recyclable." *Aluminum Association*, n.d., aluminum.org. Accessed 1 Feb. 2024.

13. "Autumn Peltier: A Long Walk for First Nations' Water Rights." *Chartered Institution of Water and Environmental Management*, n.d., ciwem.org. Accessed 1 Feb. 2024.

CHAPTER 7. WATER POLLUTION

1. Water Science School. "The Distribution of Water on, in, and above the Earth." *United States Geological Survey*, 25 Oct. 2019, usgs.gov. Accessed 1 Feb. 2024.

2. Water Science School. "The Water in You: Water and the Human Body." *United States Geological Survey*, 22 May 2019, usgs.gov. Accessed 1 Feb. 2024.

3. Water Science School, "Distribution of Water."

4. "Global WASH Fast Facts." *Centers for Disease Control and Prevention*, 31 May 2002, cdc.gov. Accessed 1 Feb. 2024.

5. "Globally, 3 Billion People at Health Risk Due to Scarce Data on Water Quality." *United Nations Environment Programme*, 19 Mar. 2021, unep.org. Accessed 1 Feb. 2024.

6. Shirin Ali. "About Half of US Water 'Too Polluted' for Swimming, Fishing or Drinking, Report Finds." *Hill*, 28 Mar. 2022, thehill.com. Accessed 1 Feb. 2024.

7. Charles E. Greer and Igor Vladimirovich. "Yellow River." *Britannica*, n.d., britannica.com. Accessed 1 Feb. 2024.

8. Yang Caini. "After the Yangtze, China Passes Law to Protect Yellow River." *Sixth Tone*, 1 Nov. 2022, sixthtone.com. Accessed 1 Feb. 2024.

9. "Jacques Yves Cousteau." *Woods Hole Oceanographic Institution*, n.d., whoi.edu. Accessed 1 Feb. 2024.

10. "Shandong Yellow River Delta National Nature Reserve." *China Daily*, 3 June 2017, chinadaily.com.cn. Accessed 1 Feb. 2024.

11. Tania Branigan. "One-Third of China's Yellow River 'Unfit for Drinking or Agriculture." *Guardian*, 25 Nov. 2008, theguardian.com. Accessed 1 Feb. 2024.

12. Jia Yan Xie, Wen Jia Tang, and Yu Hui Yang. "Fish Assemblage Changes over Half a Century in the Yellow River, China." *National Library of Medicine*, 30 Mar. 2018, ncbi.nlm.nih.gov. Accessed 1 Feb. 2024.

13. "1000 Rivers Emit Nearly 80% of Riverine Plastic Pollution into World's Oceans, Newly Published Research Shows." *Ocean Cleanup*, 30 Apr. 2021, theoceancleanup.com. Accessed 1 Feb. 2024.

14. Yan Carrière-Swallow, et al. "How Soaring Shipping Costs Raise Prices around the World." *International Monetary Fund*, 28 Mar. 2022, imf.org. Accessed 1 Feb. 2024.

15. Sharon Udasin. "Fifty Percent of US Waterways Impaired by Pollution: Report." *Hill*, 17 Mar. 2022, thehill.com. Accessed 1 Feb. 2024.

16. N. Sönnichsen. "Average Annual Number of Large Oil Spills Worldwide per Decade from 1970 to 2023." *Statista*, 30 Jan. 2023, statista.com. Accessed 1 Feb. 2024.

17. Sabrina Imbler. "In the Ocean, It's Snowing Microplastics." *New York Times*, 3 Apr. 2022, nytimes.com. Accessed 1 Feb. 2024.

18. "Restoring India's Holiest River." *United Nations Environment Programme*, 24 Mar. 2023, unep.org. Accessed 1 Feb. 2024.

19. Taniya Dutta. "Can the Ganges Be Saved from a 'Volcano' of Plastic Pollution?" *National*, 25 Nov. 2022, thenationalnews.com. Accessed 1 Feb. 2024.

CHAPTER 8. TAKING ACTION AGAINST POLLUTION

1. "Nigerian Teen Climate Activists Create Fashion from Waste to Fight Pollution." *Asharq Al-Awsat*, 25 Apr. 2021, english.aawsat.com. Accessed 1 Feb. 2024.

2. "The Recycling and Composting Accountability Act: Report." *United States Government*, 22 June 2023, govinfo.gov. Accessed 1 Feb. 2024.

3. Anne-Marie Hoeve. "The Teen Activist Who Got Plastic Banned on Bali." *Imagine5*, n.d., imagine5.com. Accessed 1 Feb. 2024.

4. Olivia Lai. "Get to Know Our EO Ambassadors, Anti-Plastic Warriors Amy and Ella Meek." *Earth.Org*, 6 Oct. 2021, earth.org. Accessed 1 Feb. 2024.

agriculture, 7, 53, 60–61, 63, 65,
    85, 94
    cows, 61, 63
    pollution, 53, 60–61, 63, 85, 94
air pollution, 6–11, 14, 16, 18–19,
    22, 25–27, 29–30, 33–38, 42–45,
    53–54, 57, 63, 73, 96, 99
Air Quality Index (AQI), 6, 25
ancient Romans, 29–30
Atacama Desert, 57, 60

baiji, 84–85
biogas, 61
birds, 17, 39, 83, 85–86
bitumen, 42, 79

Carson, Rachel, 35, 37
Central Valley, 6–7, 10
chemical pollution, 7, 10, 20, 22–23,
    26, 32, 35–37, 44, 54, 56, 59,
    61–62, 69, 72–73, 81, 83–85
    Agent Orange, 32
    DDT, 34–35, 61
    pesticides, 22–23, 26, 35, 61–62,
    65, 83
    PFAS, 56
    synthetic chemicals, 20–22, 56,
    59, 65
Chernobyl, Ukraine, 37
cholera, 30
cigarettes, 7
Clean Air Act, 36
Clean Water Act, 37
climate change, 38, 50–51, 81,
    88–89, 96–97
composting, 22, 78, 86, 95

dead zones, 70
Du, Anna, 95

electric cars, 47
extreme weather, 89

fashion industry, 57–58, 60, 64, 93
    fast fashion, 60
fertilizers, 20, 23, 26, 61, 65, 69, 72,
    78, 83
Finkbeiner, Felix, 96
fires, 6–10, 25, 29
Ford, Henry, 33–34
fossil fuels, 38, 41, 44–45, 47, 50, 53,
    57, 73, 96
    coal, 21, 33, 37–38, 41, 44–45, 50
    natural gas, 21, 41, 43–45
    oil, 6, 21, 23, 38–39, 41–43, 45,
    79, 83, 85–86
4-H clubs, 94
fracking, 44
frogs, 83

Galasso, Joseph, 95
Ganges River, 89–90
garbage, 23, 29–30, 67, 72–78, 81,
    85–87, 94
    incineration, 73
    shipping, 64, 75, 84–85
    trash picking, 78
Great Pacific Garbage Patch, 86–87
Greenagers, 94
greenhouse gases, 9, 38, 43, 45,
    48, 50–51, 53, 60–61, 63, 72,
    74–75, 81
    carbon dioxide, 13, 20, 33, 38,
    45, 63, 74
    methane, 61, 63, 74
    nitrous oxide, 22, 44, 45

human waste, 14, 22, 26, 29–30,
    67–70

industrial pollution, 31, 33–34, 38,
    53–55, 57, 60–61, 63–64, 68, 83,
    85, 89
Industrial Revolution, 31–34, 38,
    44, 66

Kyoto Protocol, 51

land pollution, 7, 14, 16, 23, 25–26,
        42, 44–46, 53–54, 60–61, 65,
        67, 70, 74, 79, 86, 88–89, 99
    landfills, 23, 58, 60, 67, 72–75,
        77–78
    soil, 15, 20, 22–23, 26, 30, 34,
        36, 54, 58, 61–63, 74, 78–79,
        94–96
light pollution, 16–17

Meek, Amy, 98
Meek, Ella, 98
miasma, 30

noise pollution, 16–17, 44, 57,
    84–85

Ocean Cleanup, 90–91
Oil Pollution Act, 38
oil spills, 38–39, 85–86
    Exxon Valdez, 38–39
organic farming, 65

Paris Agreement, 50–51
Peltier, Autumn, 79
plastic pollution, 7, 13–14, 20–24,
        26, 42, 58, 63–64, 76–78,
        85–88, 90–91, 93–95, 98–99
    microplastics, 26, 87–88, 95

recycling, 64, 76–78, 93
renewable energy, 47–50, 61, 73, 97
    nuclear, 23, 37, 49–50
    solar, 47–48, 50
    wind, 48–49

sanitation, 69–70, 76–77
Smith, Robert Angus, 33
Snow, John, 30
Sunrise Movement, 96–97
Superfund sites, 23
SustainUs, 96

Upper Fruitland, New Mexico,
    45–46, 50

water pollution, 14–16, 18, 20–23,
        25–26, 30, 35–37, 39, 44,
        53–56, 59, 61–64, 67–72, 74,
        76–79, 81–86, 88–91, 94,
        96, 99
    oceans, 13, 17, 20–21, 39, 70,
        81–82, 85–91, 95, 98
    rivers, 22, 24, 44, 60, 70, 72,
        81–85, 88–91
    wastewater, 23, 44, 54–55,
        59–60, 64, 67, 70–72
Wijsen, Isabel, 24
Wijsen, Melati, 24, 96
Wolverine Worldwide, 56

Yangtze River, 84
Yellow River, 82–83
Yosemite National Park, 5–6, 9, 11

A. W. BUCKEY

A. W. Buckey lives in Brooklyn, New York.